EXCEL

BEGINNER'S CRASH COURSE TO MICROSOFT EXCEL

MS Excel 2016, Spreadsheets, Functions & Formulas, Shortcuts

by James Clark
© 2016 James Clark
All Rights Reserved

Table of Contents

EXCEL

Table of Contents

Introduction

Chapter 1- Microsoft Excel- the beginning

Chapter 2 - Good Features Of Microsoft Excel

Chapter 3- Spreadsheets

Chapter 4 - Microsoft Excel Formulas And Functions

Chapter 5- Use Of Database In Microsoft Excel

Conclusion

Disclaimer

While all attempts have been made to verify the information provided in this book, the author does not assume any responsibility for errors, omissions, or contrary interpretations of the subject matter contained within. The information provided in this book is for educational and entertainment purposes only. The reader is responsible for his or her own actions and the author does not accept any responsibilities for any liabilities or damages, real or perceived, resulting from the use of this information.

Introduction

Behind every successful project there is always a great plan. Each and every plan is implemented in its own way and has specific requirements. Microsoft Excel help you plan well your projects. It is an essential tool when it comes to creating budgets, come up with spreadsheets for inventories, tracking your budget etc.

Microsoft Excel is not just meant for business purpose only, it has also numerous utilization at homes as it is used by different persons.

A portion of the components that you ought to comprehend about Microsoft Excel to open the probability of this application are working with spreadsheets, cell designing, different menus and toolbars, entering of information and altering the same, exploring between different spreadsheets, equation count, embedding different capacities like date/time, numerical capacities, utilizing of different diagrams that are accessible, altering, erasing remarks on cells, and so forth.

Microsoft Excel has turned into the business standard with regards to spreadsheet programming. Indeed, no other spreadsheet programming can surpass its ubiquity today. Microsoft Excel offers endless favorable circumstances to clients and throughout the years, it has been continually upgraded and extended keeping in mind the end goal to help people.

USES OF MICROSOFT EXCEL IN OUR BUSINESSES

1) Stock Control

One of the best employments of the project is for stock administration within a business. Notwithstanding the tools and materials of your occupation, Excel can monitor things. You can undoubtedly control your business stock by setting up and using straightforward formula.

 This information can then be imparted to your workers - even as they are out on the field. With Microsoft Excel, you don't need to stress over running low on things that you require.

2) Time Keeping

In the event that you have found an occupation that obliges you to figure out exactly you will be spending your time on the undertaking, Excel is the ideal tool for it. Basically, input the number of hours into Excel, you can get the aggregate hours instantly - you can then figure out the hour rate or even let you know your benefits breakdown by the hour.

3) Accounting

It is not extraordinary to utilize Excel for budgeting - where you monitor your income and costs. Also, you can utilize Excel to monitor each and every record. In an embodiment, Excel can satisfy all your business needs.

4) Invoicing

Excel can be used to create a good and organized invoice. You won't need to stress over calculating Taxes and different charges also.

5) Customer Tracking

Being a decent business individual, you ought to keep a database of your customers and the undertakings you performed for them. With Excel, you can make a spreadsheet that keeps tracks of the considerable number of subtle elements and that's only the tip of the iceberg -, for example, the employment and contact information.

This is a very profitable information that can offer you some assistance with getting more employments later on - either by the same customer or by individuals that your customer suggests for you. Just set Excel to remind you to send the email to you customers a couple of months after the occupation is finished.

6) Project Management

Use Excel to make a database ascertain the length of time of your task. With the numerous components accessible in the system, you will have no issue calculating an ideal opportunity to finish a particular employment, and in addition, the time required to finish the important errands.

7) Job Costing

Moreover, you can utilize MS Excel to compute the cost required to finish a particular assignment, and also how much the undertaking will cost you in generally.

This is a surefire approach to ensure that your venture can be finished without a soul of an over-spending plan.

Chapter 1- Microsoft Excel- the beginning

Microsoft Excel can really be used for different purposes. This is precisely why the system is intensely utilized by businesses everywhere throughout the nation. Underneath the misleadingly short-sighted configuration, Microsoft Excel is an effective system that helps your business to prosper.

Microsoft Excel is found in the Microsoft Office Appplication or Suite. It has a simple to utilize interface with various instruments that can result making a spreadsheet quick and straightforward. This joined with a capable promoting effort has made Excel a standout amongst the most prevalent programming programs on the planet.

A spreadsheet program called Multiplan was initially discharged by the organization in 1982, however, in the end, it lost the piece of the overall industry because of the arrival of Lotus 1-2-3.

On account of this, Microsoft chose to make a spreadsheet program that could viably go up against the predominance of Lotus. The main variant of Excel was presented in 1985 and was accessible on the Mac.

The principal adaptation for Windows would be discharged two years after the fact. Since Lotus didn't convey their spreadsheet project to Windows rapidly enough, Excel starts to pick up a bigger offer of the business sector.

By 1988 Excel had surpassed 1-2-3, and it is one of the elements behind the accomplishment of Microsoft as a product organization.

The most recent variant of the product is Excel 2016. An Excel document will come as .xls. Various changes can be made to the interface of the project, yet the GUI will dependably be made out of columns and cells.

Data can be put in cells which will affect the information that might be available in different cells. Notwithstanding this, Excel gives the client a lot of control over the look of cells and the data that is put in them. Both Microsoft Word and PowerPoint were intended to fit in with Excel.

The presentation of Visual Basic with Excel permitted various assignments to be robotized. Since 1993, Visual Basic has turned into an essential piece of Excel, alongside the presentation of the incorporated improvement environment.

The mechanized properties of Excel with Visual Basic has brought on various full-scale infections to be made, however, a hefty portion of them are presently hindered by standard antivirus programs.

Microsoft likewise permits clients to cripple the utilization of macros on the off chance that they decide to, and this has to a great extent dispensed with the issue.

While Microsoft Excel was not understood amid the late 1980s, it has now turned into the most broad spreadsheet programming, however, it is confronting rivalry from various organizations, most remarkably Google.

Notwithstanding this, Microsoft has become famous with the arrival of Excel, and alongside Windows, it is a standout amongst the most surely understood programming bundles on the planet.

It has excellent computation apparatuses, and it can successfully be utilized for diagramming. The product wouldn't have the strength that it has today in the event that it hadn't been for Multiplan, the antecedent that began it all.

THE BASIC ELEMENTS OF EXCEL

1) Menu - The list of items along the top of the screen; for example, file, insert, page layout etc.

2) Name box - Just underneath the Ribbon you have a white box on the left-hand side, it shows the cell reference (default A1) or if you have specified a name for a cell or range of cells, it will show that name.

3) Formula bar - Next to the Name box, also underneath the Ribbon. The formula bar shows you the contents of a selected cell, particularly useful if you want to see a calculation within a cell and not just the output of the calculation.

4) Spreadsheet - A method of spreading information across a sheet of paper. The screen represents a piece of paper with grid lines.

5) Workbook - The file you create in Excel.

6) Worksheet - A page within the Workbook. By default an Excel Workbook contains 3 worksheets; they are the viewed using tabs along the bottom.

7) Cells - The grid lines make rectangular boxes - known as cells. They are referred to using a letter and a number.

8) Columns - Cells down a spreadsheet are columns - letters.

9) Rows - Cells across the spreadsheet are rows - numbers.

MERITS OF USING MICROSOFT EXCEL

1) Similarity - Microsoft Excel clients need not stress over similarity and their capacity to send and get spreadsheets from business associates and companions. Excel is an important thing today that there is no compelling reason to change it into an alternate organization.

The similarity with other information controls programming. Excel can be utilized inside of a dominant part of other programming applications. It can be seen, embedded, and controlled effortlessly.

Regardless of the possibility that you are utilizing an outsider bookkeeping programming, you will even now have the capacity to utilize Excel in the meantime.

2) Capable customization - In spite of the fact that not many individuals know, Microsoft Excel has numerous different elements which make customization a considerable measure less demanding. This is undoubtedly incredible news for individuals who have particular programming needs.

3) Simple to utilize - You don't need to be a scientific genius to make sense of it. The straightforward usefulness and instinctual configuration are not that hard to comprehend by any stretch of the imagination.

4) Accessibility of help - Much has been composed about Microsoft Excel so it is not that hard to discover writing about it when you require help. There is additionally a few stroll through recordings, how-to articles and free instructional exercises which anybody with a tablet and web association can get to.

5) Capacity to sort out, import and investigate gigantic measures of information - In the event that you are working with monstrous measures of information which you have to import, Excel is the response for you.

It is powerful to the point that it can bolster spreadsheets with 1 million lines and a huge number of sections. It can likewise bolster multi-center processor stages for snappier figuring's.

6) Secure imparting of records to others - Many people are worried that the data they share with others over the Internet might be gotten to by other individuals. It is entirely unsettling

to discover that classified data has been shared by other individuals who shouldn't see it. With the new Microsoft Excel, this no more an issue.

7) Record size diminishment - The new packed Microsoft Excel diminishes the measure of records and it has likewise made harmed document recuperation a great deal simpler. There will be reserve funds away, and also, data transfer capacity pre-requisites. IT work force need not troubled by any means.

SOME OF THE EXCEL FORMULAS

1. SUM

Formula: = SUM A1, B1) or =SUM(6, 6) or =SUM(A1:B5)

If you want to make the addition of the two numbers then at that time this formula can be used. You can also make the addition by take cell reference.

Our first function is the SUM function. A basic step when analysing data is being able to add up or sum data and show the total of that data.

	A	B
1		
2		Numbers
3		1
4		5
5		8
6		3
7		9
8		5
9		

Say you have a list of numbers in an Excel file that you need to add up in order to generate the total.

= 1 + 5 + 8 + 3 + 9 + 5

The first way would be to add up the list of numbers using the + sign as per our arithmetic calculation, remember – link to the cell rather than manually typing in the number.

= B3 + B4 + B5 + B6 + B7 + B8

	A	B	C
1			
2		Numbers	
3		1	
4		5	
5		8	
6		3	
7		9	
8		5	
9		=B3+B4+B5+B6+B7+B8	
10			

This looks very long and complicated, with a really long list of numbers it's very easy to make a mistake; double count a number or miss one out. With a SUM function we can do this in a straight-forward way.

SUM will add or sum up the list of numbers, so does exactly the same thing, however rather than selecting each cell you can highlight the beginning and drag your mouse from the start down to the end cell and get the total in a much shorter calculation. This formula calculation and output is exactly the same as the one above however is much easier to read and much quicker to generate.

= SUM (range of cells to add up)

	A	B
1		
2		Numbers
3		1
4		5
5		8
6		3
7		9
8		5
9		=sum(B3:B8)
10		

Note here the syntax of formula.

1) The equals sign = tells Excel that you want it to do calculation – Calculation Trigger Every formula begins with an equals =, when you put an equals at the start of the cell you are telling Excel that you want it to calculate something for you.

2) The function SUM tells Excel what you want it to do – Instruction. After the equals sign you get the formula instruction, also called a FUNCTION. This is the formula or the type of calculation that you're asking it to do.

3) The brackets ( ) tells Excel you want it to do the calculation on whatever is within the brackets – Start and End of

Calculation. Next everything inside the brackets is within the calculation that Excel actions.

4) The range, notice the use of the 'colon' : to signify the range. From B3 through to B8 - B3:B8. – The Calculation

Chapter 2 - Good Features Of Microsoft Excel

Microsoft Excel 2016 is an easy to understand tool and with a lot of satisfaction. This new version has been transformed to greater extent and new features has been as well updated to make your way of working easier.

One of these new features is the new Ribbon framework is much more instinctive and inevitably numerous clients come to incline toward this new route framework.

1) THE EXCEL 2016 RIBBON MENU

The 2016 form of Excel imprints a takeoff from the old menu and toolbar arrangement of route utilized by Microsoft from the beginning of Office 95 up to 2003.

This has been supplanted with a new Ribbon framework which has been intended to require fewer mouse snaps to access generally tools. The Ribbon route frame-work comprises of a progression of Tabs which each contains Groups of tool buttons. At the point when every tab is clicked, the Ribbon changes to show an alternate arrangement of Groups.

At the base right of some of these Groups, we can see a little bolt. At the point when this is clicked, further options for the Group are uncovered. Built up clients of Microsoft Excel will probably discover these menus like "hidden" well known as they are truth be told taken straight from the past adaptations.

A convenient tip in the event that you are attempting to locate the more basic tools in this new form is to recollect the old 'right mouse click' trap.

This still works in Excel 2016 and uncovers the well-known menu, offering access to capacities, for example, duplicating, gluing and designing options.

2) THE FILE MENU

The File menu gives access to New, simply open the new record and the utilization the save option and, in addition, a rundown of the most as of late utilized archives.

Late reports can be "stuck" set up by tapping the symbol. This will forestall them being pivoted off the short rundown as more archives are opened. The number of late records showed can be changed by tapping the Options button of Excel, then picking Advanced and Display.

3) TOOLBAR ACCESS QUICKLY

This is over the Ribbon and, of course, contains a couple of the most usually utilized tools, for example, Save, Undo and Redo. Additional tools can be without much of a stretch be included by tapping the drop down menu and selecting from the rundown. Tools can likewise be expelled by just deselecting from this rundown.

4) LIVE PREVIEW

New to Excel 2016, this component empowers you to see the aftereffects of arranging options before they are connected. Floating over an arranging determination causes it to temporarily show with the proper cell.

All in all, the new format of Microsoft Excel 2016 might seem overwhelming at initially, but with a little practice, we are certain you will find that it is really far simpler to utilize beneficially than past rendition.

MICROSOFT EXCEL TOOLBARS

Toolbars are an essential part of the Microsoft Excel 2016 Application and help with enhancing your profitability and effectiveness in the application. Microsoft Excel contains a scope of toolbars that you can use for an assortment of various undertakings.

The toolbar is just a little bar that contains an assortment of caches that contain a realistic picture called an Icon. Every catch speaks to one single summon on the toolbar. To use a catch everything you need to do is to position your mouse pointer over the catch and snap once with the left button of your mouse.

The Microsoft Excel application appears, of course, two key toolbars and they are the Standard Toolbar and the Formatting Toolbar. These are the most widely recognized toolbars that you will use in Microsoft Excel and contain key orders.

After your underlying establishment of the Microsoft Excel application, you will find that these two toolbars are truth be told situated on the one line. Nonetheless, you can change their position and reposition the toolbars on the screen.

To move a toolbar, position your mouse pointer over the minimal blue four spotted bar toward the beginning of the toolbar. Hold your left mouse catch down and afterward, drag. Utilizing this strategy, you can drag the toolbar either down from its present position under the Menu bar or you can drag it to one side or right.

When you resize the toolbar to one side or right, you will see that a percentage of the symbols from either toolbar will show up and vanish. This is by configuration.

Basically, the application alters the extent of the toolbar and changes the symbols in light of those that fit on the resized toolbar.

Another awesome tool included with the toolbar is the capacity to recognize what the toolbar catch does just by holding your mouse over the symbol.

Microsoft gives an alternative in the Customize dialog box, which is found under the Tools menu to indicate both the Standard toolbar and the Formatting toolbar as a matter of course.

However to have the toolbars on one line you should deselect the Show Standard and Formatting toolbars on two lines order.

In the event that you need the two Formatting and Standard toolbars on two lines, you select the check box of the charge Show Formatting and Standard toolbars on two lines order.

There is an assortment of toolbars in the Microsoft Excel application and to show (or conceal) those toolbars you should first go to the View menu and after that pick Toolbars. When you see the menu there is some imperative information you should know about.

In the event that there is a tick beside one of the toolbar names, then this shows the toolbar is now noticeable. On the other hand, no tick shows that the toolbar is covered up.

To turn a toolbar on, snap once with your left mouse catch on the toolbar name. In the event that you need to kill the toolbar, click on the toolbar name at the end of the day.

You will find that most toolbars will be situated toward the highest point of the screen, any way you can reposition the toolbar anyplace on the screen.

Chapter 3- Spreadsheets

Spreadsheet is a concept that has come hundreds of years back. With this great invention there resulted a really easy way for arranging your data in the columns and rows. You can store anything like number, word or any kind of object in these cells. Some of the things that are mainly used in this spreadsheets are words and numbers.

This new function of spread sheeting essentially takes the manual procedures portrayed above and places them into an electronic configuration. Whilst the book-keeping industry most generally uses spreadsheets because they can be utilized as a part of any circumstance.

The sorts of businesses that use spreadsheets incorporate the science industry, aviation industry, financial services industry, common and mechanical building industry, science, instruction, research, bio-sciences, sea life science, the rundown is perpetual.

Information is characterized as far as spreadsheets as content, numbers or numerical equations. The information is what is utilized to make the electronic model and the model then appears on the screen. A model generally is an electronic rendition of a genuine circumstance.

For instance, you could show your organizations gainfulness and after that run situations in light of what cash you spent, what cash came in and when. You could even research whether an income emergency was likely with a specific situation.

The key point of interest is their capacity to have their information altered rapidly and the outcome be given quickly. Computations embraced utilizing the manual technique could have taken hours, months or even years.

Once the information is gone into a spreadsheet it can be utilized to create Graphs for archives, reports or even presentations.

For instance, you could enter your information in Microsoft Excel and after that graph creation. You could then duplicate and give the graph into a PowerPoint presentation.

DESIGNING OF SPREADSHEET IN MICROSOFT EXCEL

Designing is said to be a piece of worksheet fulfillment. It covers the elements in designing, for example,

1) Adding images currency and percent

2) Shading exchange columns data

3) Changing the data arrangement

4) Widening columns

Guidelines in Formatting and Creating Excel Charts

A chart or graph is known to be the visual representation of spreadsheet data. It regularly offers you to effectively understand the given data inside of the worksheet since are clients are offered with some assistance with effortlessly pick out the patterns and examples, which is represented in the chart. In Microsoft Excel, the sorts of graphs or charts that are regularly utilized are column chart, line graph, pie chart, bar graph and Sparkline's, which are accessible in Microsoft Excel 2016.

WORKBOOK CREATION USING MS EXCEL

In Microsoft Excel 2016, you don't need to make each worksheet yourself. There is an arrangement of preset template planned and put away in Microsoft Excel. When you open another worksheet, as a beginner to Microsoft Excel, it is somewhat startling to have a clear worksheet and you don't recognize what to do with it.

You can discover verging on each arrangement you require, receipt, charging explanation, individual month to month spending plan, deals report, time card, cost report, advance amortization etc. You could even discover more from the Microsoft on the web.

We should experience the progressions on how you could discover the templates installed on your PC. You should simply tap on Office Button and after that take after by selecting New. The New Workbook windows will show up, and you will see a rundown of exercise manual which incorporates templates that you are searching for.

The rundown of template might change starting with one Microsoft Excel then onto the next, yet in a general sense, you discover the window is the same. There are two sheets, in the left; you will discover a rundown of template classes. In the right sheets, you will see the substance of every classification, on which you will locate your new template from.

By and large, there are five noteworthy classes, specifically Blank and later, Installed Template, My Templates, New from

Existing and Microsoft Office Online, in this session, we are going to concentrate on three of them, that is Blank and Recent, Installed Template and Microsoft Office Online.

1) Clear and Recent (Blank Workbook) - is the place you begin your Microsoft Excel utilizing clear worksheet, the vast majority of the clients begin their Excel from a clear worksheet, yet for the individuals who needs help, this may not be engaging.

2) Template that is Installed - is the place the preinstalled template are put away, for a recently installed Microsoft Excel, you will discover receipt, charging proclamation, individual month to month spending plan, deals report, time card, circulatory strain tracker, cost report and advance amortization. There are more to be found in Microsoft Excel, you will discover significantly more templates at the following class:

3) Microsoft Office Online – it has distinctive subclasses that give access to various shape and templates. You could discover practically all that you require there. On the off chance that you are working together, you will

discover subclass Invoices, Inventories, Memos thus on totally valuable.

Chapter 4 - Microsoft Excel Formulas And Functions

The formulas and functions are the two fundamental essential elements in Microsoft Excel that are frequently utilized. There are many functions and formulas, which are utilized to make a spreadsheet in Microsoft Excel. It is essential that learner Excel software engineers ought to figure out how to utilize them as quickly as could be expected under the circumstances.

One of these functions is the Function LARGE. The function LARGE in Microsoft Excel gives back the kth largest worth inside of the arrangement of data. In this manner, the client can utilize this function in selecting the worth, which depends on the given relative standing.

For example, the client can totally utilize this function to give back the runner-up, third-put or even the most noteworthy score. 'LARGE (array,k)' function is to be utilized for this calculation. K is viewed as the position inside of the scope of cell or data array, which must be returned.

MICROSOFT EXCEL FORMULAS

At the point when Microsoft Excel is utilized to build huge and complex spreadsheets containing numerous worksheets, it can turn out to troublesome for somebody to explore, not to mention track or trace numbers through the spreadsheet with the majority of the qualities and formulas contained within it.

This is on account of in an expansive spreadsheet like a financial model, there can frequently be hundreds if not a great many diverse formulae down and over every page including figuring which might contain a few links to different sheets within the exercise manual.

Microsoft Excel comes with some essential usefulness to offer clients some assistance with navigating through a spreadsheets formulas.

- Using formula auditing to attract bolts to a point of reference or ward cells and double tapping to move forward and backward between same sheet references and the 'Go To' window to move forward and backward between off-sheet references.

This usefulness and the majority of the inefficiency by double tap of mouse misses the mark concerning the imprint for some clients who have vast complex spread- sheets, for example, a financial model with formulas that link to various cells or cells much further down/over the sheet or to cells on an assortment of different worksheets, or even exercise manuals.

The primary concern is that with regards to formula route the existing usefulness of Microsoft Excel is hard to utilize, inefficient and lacking in usefulness.

This regular complaint with Microsoft Excel is heard on numerous occasions by engineers, bookkeepers, administration advisors, brokers and finance experts who work with Excel spreadsheets every day.

Likely the most well-known and broadly utilized Excel include 'Formula Navigator'. They have added to a special include utilizing a re-sizeable floating window and hyperlink framework to Excel clients comprehend a formula and productively bounce to the majority of its point of reference and ward cells and ranges regardless of whether they are on an alternate worksheet or an alternate open exercise manual.

An extra history window included as a feature of the items second discharge additionally permits the client to click back to any phone beforehand took a gander at using the tool, during the present open session.

Whether the spreadsheet was composed by you or another person, 'Formula Navigator' definitely fills a need and has effectively cured the dissatisfaction of numerous Excel spreadsheet clients, helping them to rapidly comprehend the rationale in a formula and links between sheets, consequently helping to diminish facilitate debugging, errors in spreadsheet and spreadsheet auditing assistance.

Chapter 5- Use Of Database In Microsoft Excel

A database generally is what is viewed as an accumulation of data that is connected in some way. Putting away this information in a database would bode well as the organization data and the offering of an item is connected and thusly would be suitable for the database.

There are a wide range of sorts of databases accessible, for example, MySQL databases, Oracle Databases, Microsoft Access Databases, thus on however Microsoft Excel likewise has a type of a database known as a database list.

The type of the rundown is essentially the same as alternate databases as the information is under section headings in columns, however after that basic point, the Excel database goes in its own bearing.

MAKING AN EXCEED EXPECTATIONS DATABASE

Well, for one thing, there is one tenet we should dependably follow and that is one exceed expectations database for every worksheet. Any longer and you simply get yourself into heaps of inconvenience.

Truth be told in the event that you need various exceed expectations databases inside of your exercise manual basically put each exceeds expectations database onto a different worksheet.

The following thing you should follow is that your database records the first column must contain the heading of the rundown. That is the primary line that contains your field names.

Furthermore, each of the field names must be 100% one of a kind. You can't have two field names with the same name or again you will have a rundown that won't work. The following issue you should be worried about is recognizing the field names.

Exceed expectations databases have a straightforward guideline, the field names or section names must be exceptional. Presently the way you distinguish them is simple, you should simply guarantee the field names are a wide range of information sorts, design, design and so forth to whatever is left of the database in your rundown.

A standout amongst the most critical tenets you should recollect when you make an exceed expectations database is that around the line and segments of the fields and information there must be a clear line and segment.

This means you can at present have a heading at the highest point of the fields, yet there must be a clear line between the heading and the fields and along the last segment too. The clear line governs additionally applies to the base of the run down also.

When you are entering information into your rundown, each phone in each record must contain some quality regardless of the fact that it is basically clear (a clear esteem is still viewed as a worth) and every record must contain the same number of fields.

On the off chance that there is no particular information for a field you just abandon it clear and move to the following field.

Guarantee that when you are entering information into a field, you don't have spaces before the content or toward the end of the content in the field. On the off chance that you do have spaces, then what will happen is that sorting and scan for information in the rundown will be bargained and you will get startling results.

Low case and uppercase characters in the field don't influence the quests or sort orders unless you particularly tell the

Microsoft Excel application it is an issue. You can likewise utilize equations in a cell if required.

Recipes can allude to cells inside of the Excel Database List or outside of the Excel Database.

Note additionally that you can alter and organize the cells simply like some other spreadsheet, however, you need the field names must have an alternate arrangement to whatever is left of the information in the database list.

It is much prescribed that there be no other designing in the run- down aside from the field headings. This guarantees there are no erroneous conclusions by the application in the matter of what is a field heading in the exceed expectations database and what is definitely not.

FORMING AN EXCEL DOCUMENT

The Microsoft Excel restrictive formatting highlight permits the client to highlight cells that meet certain criteria. You can naturally highlight any cell whose quality meets your predetermined criteria. Contrast this with physically highlighting cells and you will see that Microsoft Excel's contingent formatting is a gigantic help.

How about we take a straightforward illustration of why it is the easy approach to go. Suppose that you needed to highlight each cell with a quality that is under 10.

The primary alternative is to physically take a gander at every cell, and if the worth is under ten, you would physically apply the red cell shading.

Presently this is fine in the event that you have a little number of cells or a great deal of spare time to do it along these lines. In any case, imagine a scenario where you have ten thousand cells to accept.

That is simply not pragmatic. All the same, in the event that you apply Excel's condition formatting highlight to those cells is simple. When you see exactly how simple it is, you too will be an enthusiast of contingent formatting.

Presently, to apply it you have to comprehend the strides that you have to take to execute it. The initial step is to indicate the arrangement of cells in your Excel exercise manual to which you need to apply the contingent formatting. The second step is to indicate the criteria to be utilized.

The third and final step is to apply it to the scope of cells to be approved. Implementing it is not hard to do, but rather it takes a period to completely see how it functions. It is not precisely intuitive. Many people can figure out how to utilize it in around 60 minutes.

However, depending on your requirements and your capacity to get new things it may be a great deal not as much as that.

REFERENCING IN MS EXCEL

In the business world, it is frequently more profitable to part your Microsoft Excel information onto numerous worksheets as opposed to packing the greater part of the information onto one single Microsoft Excel worksheet.

Hence, on the off chance that you are going to part your information over numerous exceed expectations worksheets you should have the capacity to reference that information. The names given to those references are called 3-D References.

The way you compose the Reference of 3-D is as per the following: Worksheet Name and Cell Address

You can include 3-D References in your formulas by just selecting, so as to type the reference into the cell or every cell in every worksheet as you make the formula. It is likewise conceivable to have the Reference of 3D made for you without really typing the quality in.

The initial step is to sort the equivalents sign in the cell where you need the 3-D Reference to go and after that just tap on the worksheet where the cell you need to utilize and afterward select the cell. You will now see in the Entry territory the cells 3-D Reference.

It is additionally conceivable to allude to a scope of cells using a Reference of 3-D on various worksheets or for different cells.

For instance, you might need to reference the same cell on three distinct worksheets. To do this all you would really do is basically sort for the sake of the worksheet, then press the colon key and after that sort the name of the last worksheet.

Conclusion

This book have taken you through the some basic and powerful Excel functions which you can put to the test today right now.

You can use the simple formulae in order to make sense of Excel data and do your own calculations which will dynamically update as figures are tweaked.

With the simple guidelines, you will real keys to get a handle on the basics first, then build on your knowledge. Microsoft Excel learning can be an ongoing and fulfilling process, just take it slow and build on your knowledge step by step.

www.ingramcontent.com/pod-product-compliance
Lightning Source LLC
Chambersburg PA
CBHW070419190526
45169CB00003B/1323